FLORA OF TROPICAL EAST AFRICA

DROSERACEAE

J. R. LAUNDON

(British Museum)

Insectivorous herbs. Leaves in whorls or spirally arranged. Flowers in racemes or cymes or occasionally solitary, regular, hypogynous, hermaphrodite. Sepals 4–8, imbricate, basally connate. Petals 4–8, imbricate, free, convolute. Stamens 5–20 in 1 or more whorls. Ovary superior, syncarpous, 3–5-carpellary, unilocular; styles 3–5, free or somewhat united, simple or branched; ovules numerous, on 3–5 parietal placentas or a free-basal placenta. Fruit a loculicidal capsule. Seeds small, with endosperm.

The family is represented in Africa by three genera: *Aldrovanda* L., *Drosera* L. and *Drosophyllum* Link. *Aldrovanda vesiculosa* L. is an aquatic and has been found in the Sudan, French Equatorial Africa, Portuguese East Africa and Bechuanaland Protectorate and may well occur in our area, whilst *Drosophyllum* occurs in Morocco.

DROSERA

L., Sp. Pl.: 281 (1753) & Gen. Pl., ed. 5: 136 (1754);
Diels in E.P. IV. 112: 61 (1906)

Mostly perennial herbs. Leaves in a basal rosette or in whorls, the blades provided with sticky glands which in most species are situated towards the margin of the upper surface and which trap small insects. Inflorescence 1–many-flowered, racemose or cymose. Sepals and petals usually 5, rarely 4 or 8. Stamens 5–20. Ovary of 3–5 carpels; styles 3–5. Fruit a capsule with 3–5 valves. Seeds small and numerous.

A genus of about 90 species, found in most parts of the world, but with a majority of species in Australia. *Drosera* is found throughout Africa south of the Sahara, except in South West Africa and Bechuanaland Protectorate, and is the only genus in the family known to occur in tropical East Africa.

Leaf-blades linear; peduncles usually shorter than the leaves; stipules absent; annual species . . . 1. *D. indica*

Leaf-blades orbicular, elliptic, obovate or spathulate; peduncles longer than the leaves; stipules present; perennial species:

  Plants acaulescent; leaves always in a distinct basal rosette; seeds ± ovoid, 0·3–0·6 × 0·15–0·2 mm.:

    Peduncles glandular; petiole narrow throughout its length and broadening abruptly into the suborbicular blade; flowers usually white or pink . . . . 2. *D. burkeana*

Peduncles pilose ; petiole broadening gradually into the elliptic or obovate blade ; flowers reddish . . . . . . 3. *D. pilosa*

Plants normally caulescent, but rarely ± acaulescent by reduction ; seeds fusiform, 0·7–0·9 × 0·2 mm. :

Petiole 2–5 times length of blade, very slender, glabrous or very sparsely pilose on both surfaces, erect in age . . . . . 4. *D. affinis*

Petiole 1–2 (–3) times length of blade, slender, usually pilose on under surface, reflexed in age . . . . 5. *D. madagascariensis*

1. **D. indica** *L.*, Sp. Pl. 282 (1753) ; Cat. Welw. Afr. Pl. 1 : 330 (1896) ; Diels in E.P. IV. 112 : 77, Fig. 29 (1906) ; Fl. Pl. Sudan 1 : 82 (1951) ; Taton in F.C.B. 2 : 551 (1951). Type : " India " [actually Ceylon], a drawing in Herb. Hermann 5 : t. 227 (BM, holo. !)

Caulescent annual. Stem 3–50 cm. long, glandular-pubescent, leafy. Leaves spirally arranged ; blade linear, 10–100 mm. long, 0·5–3 mm. broad, glandular ; petiole 1–15 mm. long, glandular-pubescent. Stipules absent. Peduncles lateral, usually extra-axillary, 5–150 mm. long, glandular-pubescent ; inflorescence 3–20-flowered ; pedicels 3–20 mm. long, glandular-pubescent. Sepals 5, ± lanceolate, 3–5 mm. long, 1·0–1·5 mm. broad, glandular-pubescent. Petals 5, obovate, 6–8 mm. long, 3–6 mm. broad, pink to purple. Stamens 5, 3–4 mm. long. Ovary subglobose ; styles 3, bifid nearly to the base. Seeds ovoid, apiculate, 0·4–0·5 mm. long, 0·3 mm. broad ; testa reticulate with longitudinal and transverse ridges. Fig. 1/11.

UGANDA. Acholi District : Kilak Hill, NW. of Gulu, 19 Nov. 1941, *A. S. Thomas* 4048 !, & Nov. 1941, *Eggeling* 4698 !

TANGANYIKA. Tanga District : Lwengera Valley, about 6·5 km. E. of Korogwe, 20 July 1953, *Drummond & Hemsley* 3369 ! ; Rufiji District : Mafia Island, Kilindoni, 6 Aug. 1936, *Vesey-FitzGerald* 5214 ! ; Songea District : about 8 km. W. of Songea, 18 June 1956, *Milne-Redhead & Taylor* 9807B !

DISTR. **U**1 ; **T**3, 6, 8 ; chiefly lowland tropical Africa from Senegal to Portuguese East Africa ; also in Madagascar, Asia from India to Japan, New Guinea and Australia

HAB. Boggy places, and on seasonally wet acid rocks in open spaces in *Brachystegia* woodland ; becoming a weed of damp places in derelict cultivated ground ; 0–1100 m.

2. **D. burkeana** *Planch.* in Ann. Sc. Nat., sér. 3, Bot. 9 : 192 (1848) ; F.T.A. 2 : 402 (1871) ; Cat. Welw. Afr. Pl. 1 : 330 (1896) ; Diels in E.P. IV. 112 : 88 (1906) ; Taton in F.C.B. 2 : 552 (1951) ; Exell & Laundon in Bol. Soc. Brot., sér. 2, 30 : 217, 218, pl. 2 (1956). Type : South Africa, Transvaal, Macalisberg, *Burke* (K, holo. !)

Acaulescent perennial. Leaves in a basal rosette ; blade suborbicular, 2–10 mm. long, 2–9 mm. broad, glandular on and around the margin of the upper surface, glabrous on the lower surface ; petiole 2–20 mm. long, narrow throughout and broadening abruptly into the blade, glabrous or pilose. Stipules 3 mm. long, connate at the base ; apex lacerated. Peduncles 1–4, 4–30 cm. long, arising laterally from the rosette then curving to become erect, canaliculate, glandular ; inflorescence racemose, often secund, 2–12-flowered ; pedicels 2–12 mm. long, glandular ; bracts narrowly obovate, 1–2 mm. long, glandular or glabrous. Sepals 5, ± elliptic, 4–5 mm. long, 2 mm. broad, acute or obtuse, irregularly serrulate at the apex, glandular. Petals 5, 5–7 mm. long, 3–4 mm. broad, white or pink. Stamens 5 ; filaments 4 mm. long. Ovary subglobose, glabrous ; styles 3, bifid nearly to the base.

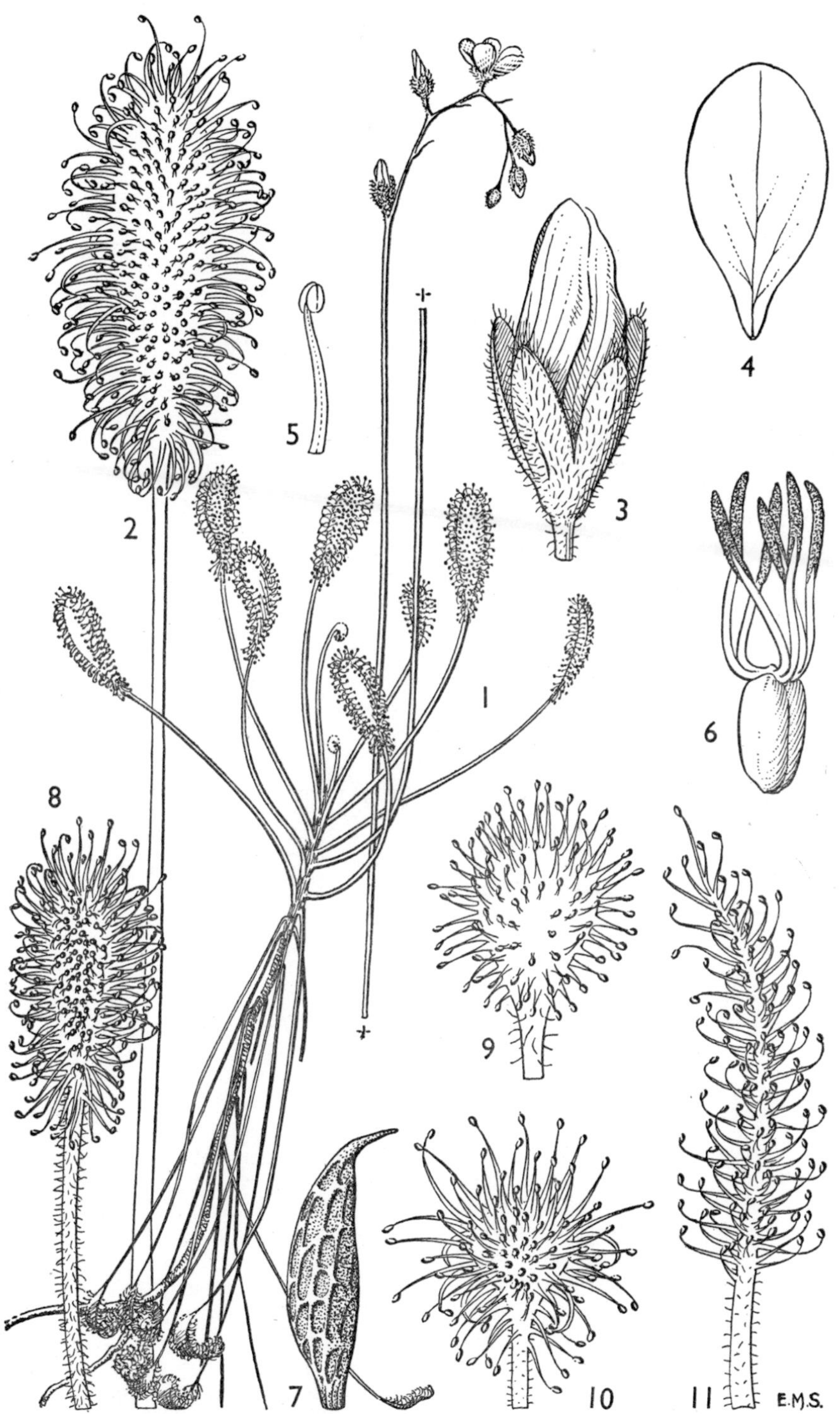

FIG. 1. *DROSERA AFFINIS*—**1**, plant in flower, × 1; **2**, leaf, × 4; **3**, flower-bud, × 8; **4**, petal, × 8; **5**, stamen, × 8; **6**, pistil, × 12; **7**, seed, × 60; *D. MADAGASCARIENSIS*—**8**, upper part of leaf × **4**; *D. PILOSA*—**9**, upper part of leaf, × 4; *D. BURKEANA*—**10**, upper part of leaf, × 4; *D. INDICA*—**11**, upper part of leaf, × 4. 1–7, from *Milne-Redhead & Taylor* 10485; 8, from *Milne-Redhead & Taylor* 8490; 9, from *Stolz* 1885; 10, from *Milne-Redhead & Taylor* 10829; 11, from *Milne-Redhead & Taylor* 9807B.

Seeds ovoid, 0·3–0·4 mm. long, 0·15–0·2 mm. broad, black ; testa smooth. Fig. 1/10, p. 3.

UGANDA. Masaka District : Lake Nabugabo, Aug. 1935, *Chandler* 1348 !
TANGANYIKA. Bukoba District : Biharamulo road, 5 km. from Bukoba, Aug. 1931, *Haarer* 2054 ! ; Rungwe District : Kyimbila district, 8 Sept. 1910, *Stolz* 252 ! ; Songea District : 8 km. W. of Songea, 18 June 1956, *Milne-Redhead & Taylor* 10829 ! & 10829A !
DISTR. **U4** ; **T1**, 7, 8 ; Angola, Belgian Congo (Katanga), Nyasaland, Northern and Southern Rhodesia, Portuguese East Africa, Transvaal & Madagascar
HAB. Boggy places on poor soils in short grassland and in open spaces in *Brachystegia* woodland ; 900–1400 m.

3. **D. pilosa** *Exell & Laundon* in Bol. Soc. Brot., sér. 2, 30 : 213 (1956). Type : Tanganyika, Rungwe District, near Kyimbila, *Stolz* 1885 (BM, holo. !, K, iso. !)

Acaulescent perennial. Leaves in a basal rosette ; blade oval or obovate, 4–13 mm. long and 2–8 mm. broad, glandular on and around the margins above, pilose beneath ; petiole 3–15 mm. long, broadening out gradually into the blade, glabrous on the upper surface and pilose beneath. Stipules 3 mm. long, connate at the base ; apex lacerated. Peduncles 1–2, 2–16 cm. long, arising laterally from the rosette and curving to become erect, canaliculate, pilose throughout their lengths; inflorescence racemose, 2–6-flowered ; pedicels 1–7 mm. long, pilose ; bracts elliptic, 2–3 mm. long. Sepals 5, ± elliptic and acute, 3–6 mm. long, 1·5–2·5 mm. broad, pilose. Petals 5, about 6 mm. long, red or reddish-purple. Stamens 5 ; filaments 4 mm. long. Ovary subglobose, 2 mm. long, 1·5 mm. broad ; styles 3, 2 mm. long, bifid nearly to the base. Seeds ovoid, 0·3–0·5 mm. long, 0·2 mm. broad, black ; testa smooth. Fig. 1/9, p. 3.

KENYA. Nakuru District : Mau Forest, Bondui, 22 Jan. 1946, *Bally* 4940 !
TANGANYIKA. Rungwe District : N. slopes Mt. Rungwe, 14 Mar. 1932, *St. Clair-Thompson* 972 !
DISTR. **K3** ; **T7** ; British Cameroons
HAB. Open marshy and rocky places in upland rain-forest, upland grassland, upland evergreen bushland ; 2200–2700 m.

SYN. *Drosera* sp., W.F.K. : 14 (1948)
[*D. burkeana* sensu Keay in F.W.T.A., ed. 2, 1 : 121 (1954), *non* Planch.]

4. **D. affinis** [*Welw. ex*] *Oliv.*, F.T.A. 2 : 402 (1871) ; Cat. Welw. Afr. Pl. 1 : 330 (1896) ; Diels in E.P. IV. 112 : 88 (1906). Type : Angola, Huila, near Lopolo, *Welwitsch* 1183 (LISU, lecto., BM, COI, K, isolecto. !)

Caulescent perennial. Stem 1–23 cm. long. Leaves spirally arranged, mostly erect, but those on the lower part of the stem sometimes descending ; blade narrowly oblanceolate or narrowly oblanceolate-oblong, 3–30 mm. long and 2–5 mm. broad, glandular around the margin of the upper surface, glabrous or sparsely pilose on the lower surface ; petiole 5–70 mm. long, very slender, glabrous or very sparsely pilose. Stipules 3–13 mm. long ; apex lacerated. Peduncles 1–4, 8–30 cm. long, arising laterally from the stem, then curving to become erect, canaliculate, glabrous ; inflorescence racemose, 3–13-flowered ; pedicels 2–10 mm. long, glabrous or sparsely pilose ; bracts linear to elliptic, 3–5 mm. long, caducous. Sepals 5, oblong-lanceolate, 5–7 mm. long, 1·5–2 mm. broad, pubescent or pilose. Petals 5, 5–8 mm. long, white or purple. Stamens 5 ; filaments 5 mm. long. Ovary subglobose, glabrous ; styles 3, 2·5 mm. long, bifid nearly to the base. Seeds fusiform, 0·7–0·9 mm. long, 0·2 mm. broad, brownish-black ; testa reticulate with longitudinal and transverse ridges. Fig. 1/1–7, p. 3.

TANGANYIKA. Lushoto District : Magamba, near Lushoto, Jan. 1954, *Eggeling* 6785 ! ; Njombe District : Uwemba, May 1953, *Eggeling* 6534 ! ; Songea District : 1·5 km. E. of Songea, 3 June 1956, *Milne-Redhead & Taylor* 10485 !

DISTR. **T3**, 7, 8 ; Angola, Belgian Congo (Katanga), Nyasaland, Northern and Southern Rhodesia and Portuguese East Africa

HAB. Boggy places and seepage areas on poor soil in short grassland and in open spaces in *Brachystegia* woodland ; 900–2200 m.

SYN. *D. flexicaulis* [Welw. ex] Oliv., F.T.A. 2 : 403 (1871) ; Cat. Welw. Afr. Pl. : 331 (1896) ; Diels in E.P. IV. 112 : 98 (1906) ; Taton in F.C.B. 2 : 554 (1951). Type : Angola, Huila, Morro de Lopolo, *Welwitsch* 1181 (LISU, lecto., BM, COI, K, isolecto. !)

NOTE. This species is frequently confused with *D. madagascariensis* DC., with which it often grows. Thus one commonly finds herbarium material consisting of a mixture of the two species. Milne-Redhead and Taylor studied the two species growing together near Songea, and noted that *D. affinis* occupied " rather wetter places " than *D. madagascariensis* and had longer petioles and leaf-blades and the blades were " paler to more flesh-coloured ; petioles less red ; few and shorter glands on reddish calyx ; petals mauve ; calyx more or less erect in fruit ; leaves not reflexed in age." *D. madagascariensis* on the other hand had " leaves pale red with bright red petioles . . . reflexed in age . . . petals pinkish-mauve and calyx spreading in fruit."

Forms with slender flexuous stems and shorter leaves were described by Oliver in F.T.A. 2 : 403 (1871) as *D. flexicaulis*. However, a whole range of intermediates between *D. flexicaulis* and *D. affinis* has since been found and it is no longer possible to keep them apart. Diels in E.P. IV. 112 : 82 (1906) describes the seeds of *D. flexicaulis* as fusiform and those of *D. affinis* as ovoid, but on examination the seeds were found to be fusiform in the type-specimens of both.

5. **D. madagascariensis** *DC.*, Prodr. 1 : 318 (1824) ; Diels in E.P. IV. 112 : 98 (1906) ; Taton in F.C.B. 2 : 554 (1951). Type : Madagascar, locality and collector unknown (G–DC, holo., K, photo. !)

Caulescent perennial, but sometimes apparently acaulescent by reduction. Stem to 25 cm. long. Leaves spirally arranged, erect at the top of the stem, but reflexed lower down ; blade elliptic, obovate or spathulate, 5–15 mm. long and 2–5 mm. broad, glandular, especially towards the margins on the upper surface and sparsely pilose on the lower surface ; petiole 10–30 mm. long and 0·3–1 mm. broad, glabrous or pilose above, usually pilose (rarely glabrous) below. Stipules oblong, up to 5 mm. long and 1 mm. broad, lacerated at the apex. Peduncles 1–2, 9–55 cm. long, arising laterally below the top of the stem, then curving to become erect, glabrous, glandular or pilose ; inflorescence 2–14-flowered ; pedicels 1–10 mm. long, glandular or pilose ; bracts linear, 3 mm. long, sparsely pilose, caducous. Sepals 5, elliptic or oblong-lanceolate, 4–7 mm. long, 1–2 mm. broad, glandular-pilose. Petals 5, 5–8 mm. long, pink or purple. Stamens 5 ; filaments 5 mm. long. Ovary subglobose ; styles 3, 2 mm. long, bifid nearly to the base. Seeds fusiform, 0·7–0·9 mm. long, 0·2 mm. broad, brownish-black ; testa reticulate, with longitudinal and transverse ridges. Fig. 1/8, p. 3.

UGANDA. Kigezi District : Kachwekano Farm, June 1951, *Purseglove* 3633 ! ; Masaka District : E. side of Lake Nabugabo, 6 Oct. 1953, *Drummond & Hemsley* 4651 ! ; Mengo District : Namanve swamp, July 1932, *Eggeling* 459 *in F.D.* 795 !

KENYA. Uasin Gishu District : Sergoit R., Soy Pool, 11 Jan. 1948, *Bickford in Bally* 6259 !

TANGANYIKA. Bukoba District : Biharamulo road 5 km. [from Bukoba], Aug. 1931, *Haarer* 2055 ! ; Rungwe District : Kiwira R., May 1953, *Eggeling* 6519 ! ; Songea District : about 12 km. W. of Songea, 5 Feb. 1956, *Milne-Redhead & Taylor* 8490 !

DISTR. **U2**, 4 ; **K3** ; **T1**, 4, 7, 8 ; throughout tropical and subtropical Africa from French Sudan and French Guinea to Natal and Cape Province (Pondoland) ; also in Madagascar

HAB. Boggy places and seepage areas in short grassland and in open spaces in *Brachystegia* woodland ; rarely on wet rocks ; 900–2300 m.

SYN. [*D. ramentacea* sensu auct., pro parte, *non* Burch. ex DC.]
D. congolana Taton in B.J.B.B. 17 : 310 (1945) ; F.C.B. 2 : 552 (1951). Type : Belgian Congo, Bas-Congo, *Bequaert* 7191 (BR, holo. !)

VARIATION. Specimens from Bas-Congo (Belgian Congo), Mt. Mlanje (Nyasaland), the Transvaal and Natal usually have very short stems and comparatively few reflexed leaves. Such plants have often been confused with truly acaulescent species (i.e. *D. burkeana* Planch., *D. natalensis* Diels) or described as new to science (i.e. *D. congolana* Taton). It is not known whether these forms are due to genetic differences or differences in environment. The extreme form of this type, with long petioles, was named var. *major* Burtt-Davy & Greenway by Burtt-Davy in Man. Fl. Pl. Transv. 1 : 39 (1926).

INDEX TO DROSERACEAE